SUPER-STRONG ANIMALS

BLACK RABBIT BOOKS

JOANNE MATTERN

TABLE OF CONTENTS

1

The Bald Eagle

A bald eagle is a great hunter. This bird uses its super strength to catch its prey. The eagle has large claws on its feet. These talons are as sharp as razor blades.

Eagles hunt for animals as they fly. They have good eyesight. The bird spots something to eat. It swoops down. It grabs the animal in its sharp talons.

A bald eagle can lift more than four times its body weight. They eat fish, ducks, rabbits, snakes, and turtles.

Did You Know?
The grip of an eagle is 10 times stronger than a person's.

2

The Grizzly Bear

Grizzly bears can grow to be 10 feet (3 meters) tall. They can weigh up to 1,700 pounds (771 kilograms). A bear can lift 500 pounds (227 kg) with one paw. It would take up to five average humans to hold that much weight.

These bears eat many things. They eat meat, fish, fruit, and plants. A grizzly can even kill and eat a large deer.

Think About It

Today, grizzlies only live in a small part of North America. Why don't they live in other places?

3

The Komodo Dragon

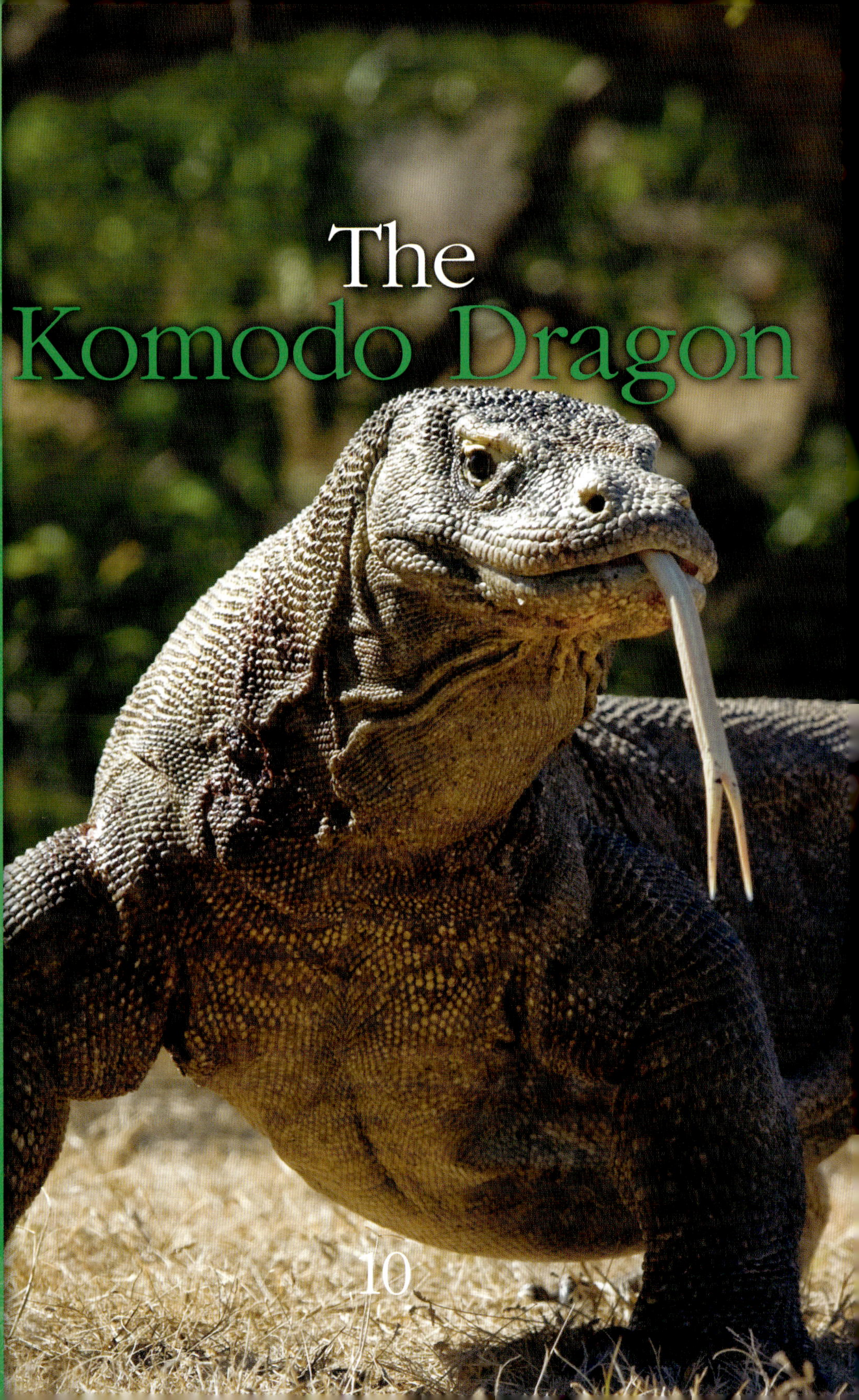

You would not win a fight with a Komodo dragon. These reptiles are super strong. They are super deadly too.

A Komodo dragon attacks animals with its sharp claws and teeth. Then it shakes them hard. This lizard eats pigs, goats, deer, and horses. It is strong enough to kill a water buffalo.

This lizard will eat anything—even other Komodo dragons! They have thousands of tiny bones under their skin. These bones are like armor. It keeps them safe during fights.

Did You Know?

This lizard has a deadly bite. Its mouth is full of **venom** and **bacteria**. If a bite doesn't kill, the animal will die soon anyway.

4

The Gorilla

A gorilla will always win a weight-lifting contest against a person. A gorilla can lift 4,000 pounds (1,810 kg). A very strong man can only lift about 885 pounds (401 kg).

Gorillas eat leaves and roots. They also eat snails and bugs. A male can weigh up to 500 pounds (227 kg). These animals have large muscles. They can rip down banana trees to get the fruit. They can also easily snap bamboo plants with their hands.

Did You Know?
Gorillas are smart, too. They use tools and can learn sign language.

5

The Leafcutter Ant

Can a small bug be super-strong? Yes! The leafcutter ant does amazing things.

This ant can carry 10 to 50 times its own weight. That is like carrying thousands of pounds. Here's something even more amazing. The ants do not carry leaves with their legs. They carry them in their jaws.

Leafcutter ants carry leaves back to their nest. They use the leaves to grow fungus. This is the ants' only food.

Think About It

How do you think leafcutter ants learned how to grow food?

6

The Dung Beetle

The grizzly is not the strongest animal in the world. Neither is the gorilla. The strongest animal is a small bug. Some dung beetles can push more than 1,000 times their body weight.

What does a dung beetle push? Poop! These bugs got their name because they eat poop, or dung.

When an animal poops, it leaves little bits of food behind. The dung beetle eats this food. Beetles use their super strength to roll balls of poop into their nest.

Did You Know?
Dung beetles lay their eggs in balls of poop so their babies have food.

MORE TO EXPLORE

WHERE IN THE WORLD?

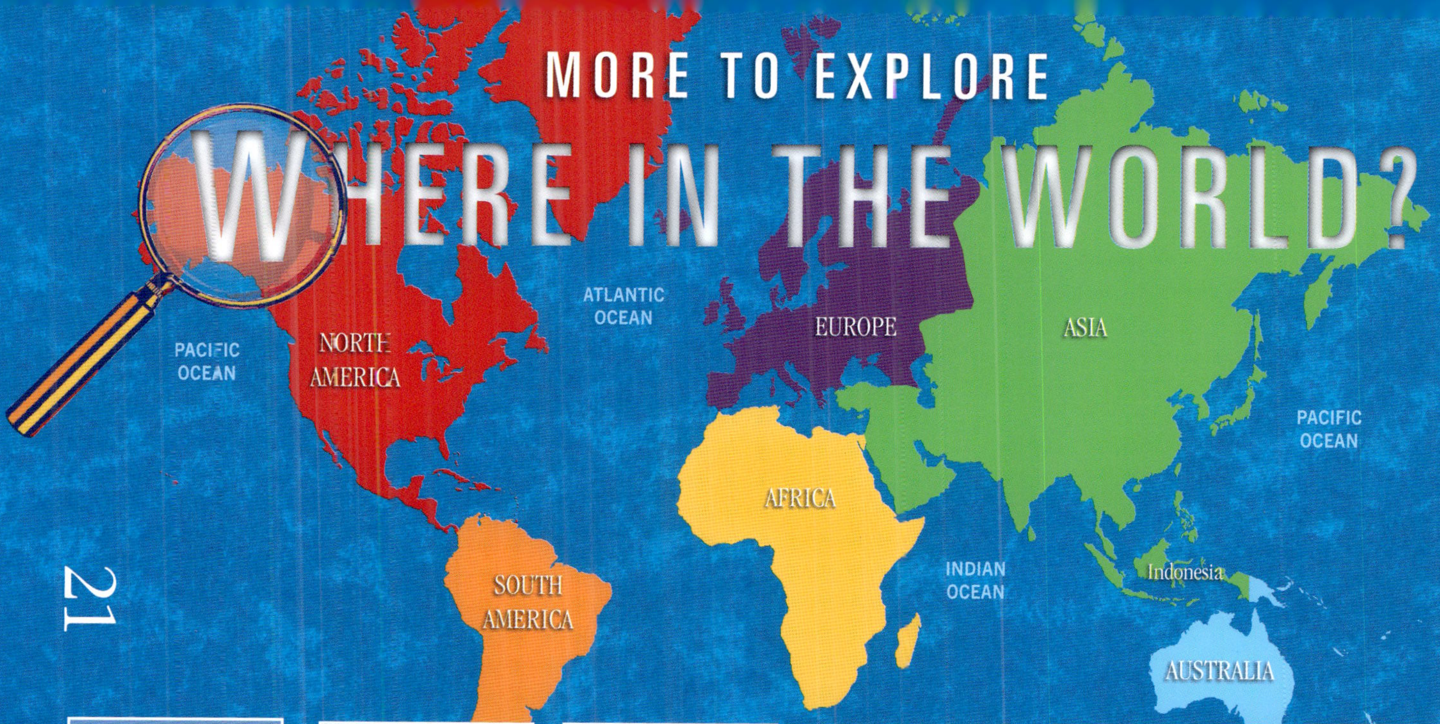

BALD EAGLES
North America

GRIZZLY BEARS
Northwestern North America

KOMODO DRAGONS
Indonesia

GORILLAS
Central Africa

LEAFCUTTER ANTS
North America and South America

DUNG BEETLES
All continents except Antarctica

MORE TO EXPLORE

FANTASTIC FACTS

Bald eagle nests can be 8 feet (2.4 m) wide. They can be up to 12 feet high (3.7 m).

A grizzly's front claws can be 4 inches (10 cm) long.

A Komodo dragon can kill with one hit of its heavy tail.

Gorillas live in big groups. The oldest male is the leader.

Leafcutter ant nests have thousands of rooms and tunnels.

Male dung beetles have horns. They hit each other with them when they fight.

MORE TO EXPLORE

COOL COMPARISONS

Most of the animals in this book are much stronger than humans. Let's compare.

A grizzly bear can lift 0.8 times its body weight.

A bald eagle can lift about 4 times its body weight.

Some humans can lift 5 times their weight.

Gorillas can lift 27 times their body weight.

Leaf-cutter ants can lift 50 times their body weight.

The dung beetle can push 1,141 times its weight.

MORE TO EXPLORE

RESOURCES

Glossary

bacteria (bak-TEER-ee-uh) A small living thing.

fungus (FUN-gus) A living thing, similar to a plant that has no flowers, that lives on dead or decaying things.

prey (PRAY) An animal hunted or killed for food.

talon (TAH-luhn) The sharp claw on the foot of some birds.

venom (VEH-nuhm) A poison made by animals used to kill or injure.

Read More

Austen, Lily. *Strongest Animals.* Minneapolis: Jump!, Inc., 2025.

Mariño, Soledad Romero. *Superpowered Animals.* New York: Phaidon Publishing, 2023.

Index

TOP RANK is published by Black Rabbit Books, P.O. Box 227, Mankato, MN, 56002. •

• Top Rank is an imprint of Black Rabbit Books. • Edited by Alissa Thielges • Designed by Danny Nanos • Photographs © Dreamstime/Elliot Hurwitt, 13, 23, Sekarb, cover, Thor Jorgen Udvang, cover; Getty Images/guentergunl, 14, 21, michaklootwijk, 16, 21, 23; Shutterstock/Eric Isselee, 2, FloridaStock, 4, 21, 23, Gallinago_media, 5, GUDKOV ANDREY, 10, 12, 21, Henk Bogaard, 19, 20, 21, 23, kkaplin, 11, Manick Photo, 23, matteolo, 9, 21, photomaster, 5, Scott E Read, 6–7, 8, 23, Stephen Meese, 17, Vinicius R. Souza, 18, Vitalii Hulai, 20, Yair Leibovich, 15 • Printed in China

Library of Congress Cataloging-in-Publication Data: Names: Mattern, Joanne, 1963- author. | Title: Super-strong animals / by Joanne Mattern. | Description: Mankato, MN : Top Rank is an imprint of Black Rabbit Books, [2025] | Series: Super-incredible animals | Includes bibliographical references and index. | Audience: Ages 8–11 | Audience: Grades 2–3 | Identifiers: LCCN 2024017535 | ISBN 9781645823834 (library binding) | ISBN 9781645824053 (paperback) | ISBN 9781645824275 (ebook) | Subjects: LCSH: Animals—Miscellanea—Juvenile literature. | Physiology—Juvenile literature. | Muscle strength—Juvenile literature. | Classification: LCC QL49 .M4353 2025 (print) | LCC QL49 (ebook) | DDC 591—dc23/eng/20240517 | LC record available at https://lccn.loc.gov/2024017535